THE DISASTER SURVIVAL HANDBOOK

A DISASTER SURVIVAL GUIDE FOR MAN-MADE AND NATURAL DISASTERS

SAM FURY

Illustrated by

DIANA MANGOBA & NEIL GERMIO

www.SurvivalFitnessPlan.com

WARNINGS AND DISCLAIMERS

CONTENTS

INTRODUCTION

The Disaster Survival Handbook is a no-nonsense reference book about what to do to give yourself (and those around you) the best chance of survival in the case of various natural and non-natural disasters.

It is presented in two parts.

Part 1 provides you with information for surviving specific disasters. These are split into two categories: natural and man-made. Disasters are listed alphabetically within each category.

Part 2 is a beginner's guide to prepping.

Prepping is "the practice of making active preparations for a possible catastrophic disaster or emergency" (Google Dictionary).

This section gives you enough information to prepare to survive a short-term "grid-down" scenario (about three weeks). It will also provide you with a good understanding of the areas you need to research further if you want to prep for longer-term scenarios.

SURVIVING DISASTERS

GENERAL DISASTER RESPONSE ACTION PLAN

Use this action plan for any disaster situation where you do not have a specific plan already in place. Skip any step that will put you in danger.

- Remain calm. This is easier said than done, but necessary.
- Check for immediate hazards before moving.
- Apply critical first aid ("critical" meaning that the person will die without immediate attention).
- Put on practical/protective clothing (shoes, long pants, etc.).
- Get your bug-out bag (BOB). Carry it with you everywhere until safe (see the *Bug Out Bags* chapter).
- Monitor the media (use the radio from your BOB).
- Appoint a leader.
- Ensure the safety of family members.
- Secure your home. Turn off water, electricity, and natural gas if needed. Clean up other hazards, such as broken glass, if possible. Secure entry points from possible looters.
- Devise a plan and act on it.
- Seek help/help others (neighbors, extended family, etc.).
- Evacuate if advised to do so.
- Improve morale by increasing comfort (food, water, warmth, etc.), keeping busy, staying positive, and creating familiarity for children and pets (e.g., with favorite toys).

AVALANCHE

An avalanche is when a mass of snow slides down a steep slope.

When in avalanche country, travel with a friend, pay attention to the weather, and carry a rescue beacon. There is a risk of avalanches in any snowy, mountainous area. They are more likely to occur:

- After a rise in temperature.
- After rain.
- In deep, snow-filled gullies.
- In the afternoon if the morning has been sunny.
- On angles of 30° to 45°.
- On the side that faces away from the wind (leeward side).
- On snow-covered convex slopes.
- Within 24 hours of snowfall lasting 2 or more hours, especially in low temperatures.

When you are crossing avalanche territory, take the following safety precautions:

- After midday, keep to slopes that have already been exposed to the sun.
- Before midday, travel in shaded areas.
- Avoid small gullies and valleys with steep side walls.
- Carry an avalanche probe and a beacon.
- Stick to ridges and high ground above avalanche paths.

If there is a group of you:

- Keep at least 20 meters apart.
- Rope together and use belays, except when skiing.
- Ski down any slopes one at a time.

Surviving an Avalanche

The action to take to survive an avalanche depends where you are in relation to it. If it starts below your feet, get upslope of any cracks in the snow. When you are below it, move to the closest side out of its path.

If you can't avoid it, get rid of all access weight and grab something solid, such as a tree. Do not abandon your ski pole or communication devices.

When there is nothing to grab onto, use the freestyle swimming stroke to stay on top of the snow. If you are unable to stay on top, put your hands in front of your nose and mouth to create an air pocket.

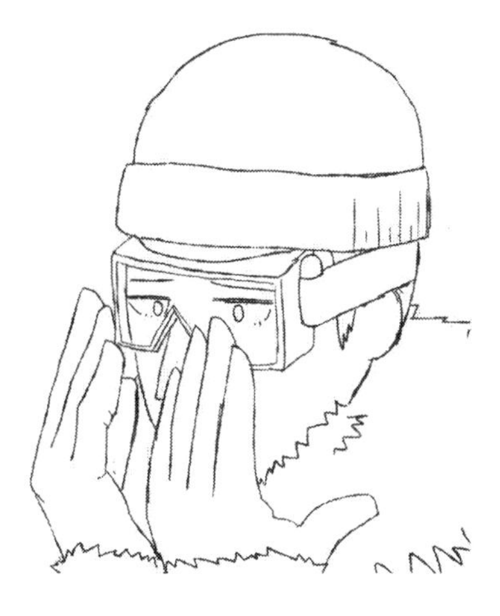

As soon as you stop, make as big an area as possible while trying to reach the surface. Use your ski pole to poke around and find open air. To figure out which way is up (to the surface), spit and go in the opposite direction from the one in which the spit falls.

BLIZZARD

A blizzard is a severe snowstorm. In the event of a blizzard, take shelter until it passes. If there is no shelter, build one.

Prepare your home for a blizzard by storing the following:

- Rock salt for walkways.
- Sand for traction.
- Shovels to clear the snow.
- Wood to burn.
- Warm clothes.

Always avoid driving if there is a possibility of a blizzard, and keep a few emergency blankets in your car just in case.

If you do get caught in a blizzard while driving, pull over and wait out the storm. Use seat covers and floor mats as emergency insulation. If you have enough fuel, you can run the engine for warmth. Cover the engine to minimize heat loss, and make sure the exhaust pipe is clear. Only run the engine for 10 minutes per hour to prevent carbon monoxide poisoning and to conserve the battery. If you start to feel drowsy, stop the engine and open a window.

Keep the hazard lights off to conserve the battery, but leave the interior dome light on at night.

If the snow is piling up, get out and build a snow shelter to avoid being trapped in your car. Only move away from your vehicle or shelter if:

- Help is accessible within a reasonable distance.
- You have visibility and conditions are safe.
- You have appropriate clothing.

Hang something bright on your shelter so you and others can find it.

How to Build a Snow Trench

A snow trench is an easy-to-build shelter that can accommodate several people if made large enough.

Dig a trench long and wide enough for you to sleep in. Use what you dig out to build sides. Face it so the wind hits the long sides and make the entrance at the lower end.

Make a roof out of sticks and vegetation, material, or compacted snow bricks leaning against each other. Fill any gaps with snow and insulate the ground with dry vegetation or whatever you have.

Keep your digging tool next to you in case the shelter collapses.

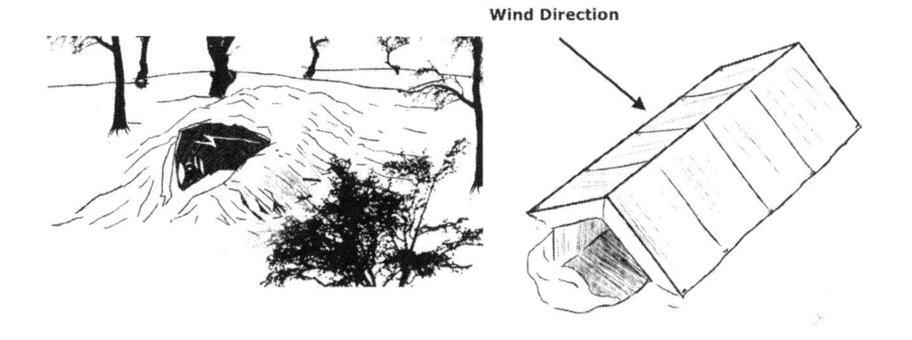

DANGEROUS ANIMALS

This chapter covers what to do if attacked by a select few animals, but the advice in it can be adapted for others.

As a general rule, leave animals alone. Keep out of their territory and away from their children. Do not feed, antagonize, or surprise them.

If you see a baby animal alone, its mother is probably close by and will be protective. Back away.

Most animals are afraid of fire and loud noises. If you use these things to scare them off, make sure you give them a way to escape. Animals are especially dangerous when they feel threatened or vulnerable.

To protect yourself from smaller animals, keep your hands and feet out of places you can't see. Use sticks to turn over logs, rocks, and other things in the wild.

If you come across a large animal, keep calm, freeze, then slowly back away. Do not make any sudden movements. Most larger animals have poor turning circles. If one charges you, run in a zigzag pattern and/or move out of the way at the last minute.

Climbing a tree to get away from a predator is a last resort. The animal may wait for you.

Alligators and Crocodiles

Alligators and crocodiles are found all over the world, in lakes, rivers, wetlands, and some coastal areas. They are technically different, but the actions you should take if you come across one are the same in either case.

When in their territory, stay out of the water. Don't even dangle your limbs over the side of a boat or riverbank.

Trying to catch an alligator (or crocodile) is hardly ever a good idea, but if you need to do it, wait until the animal is on dry land. Get on its back and force its head and jaws down by applying downward pressure on its neck. Cover its eyes to calm it.

If you are attacked, go for its eyes and nose. If it grabs your limb, tapping on its snout may cause it to open its mouth and release you.

Bears

There are several species of bear, and they live in wilderness areas all over the world. Occasionally, they may also wander into the suburbs for food.

When in the wilderness, carry bear spray and a whistle, and don't set up camp where there are bear tracks or scat.

Bears are mostly attracted to food, so do not leave it lying around near where you plan to sleep. Hang it out of reach if you can. The smell of food is as attractive as the food itself, so do not sleep in the same clothes you cook in.

If you see a bear, stop, be quiet, and observe it. Wait at least 30 minutes after it is out of sight before continuing. Alternatively, take a different route and give the bear a wide berth. Make noise as you walk the new route. If you're in a car, stay in it, and keep the windows up.

If the animal sees you, prepare your bear spray. Do not try to run, climb a tree, or even turn your back on it. Instead, slowly walk sideways to safety while watching it.

If the bear attacks you, strike back with anything you can, and aim for its eyes and/or snout.

A black bear is the least likely to attack. Getting "large" and making lots of noise may scare it away.

If you encounter a brown bear, play dead. Lay on your stomach and cross your hands behind your neck. If it doesn't stop attacking you after a few seconds, fight for your life.

Polar bears will hunt humans if they are hungry enough. Fight for your life.

Bees

Unless you are allergic, bees are not usually a problem. As with most animals, if you leave them alone, they'll leave you alone. However, if you accidentally disturb a hive, they may swarm you. This is dangerous even if you don't have an allergy.

In that case, don't try to swat them. Run indoors instead. When in the wilderness, run through bushes or high weeds for cover. Getting under water can help, but the bees may wait for you to resurface.

Most species will chase you up to 50 meters, but some may go as far as 150 meters.

If you get stung, remove the stinger as soon as possible by raking your fingernail, credit card or something similar across it in a sideways motion.

Boars

Wild boars are extremely tough. They are not likely to charge, but if they do it is most likely to be at dusk, dawn, and in winter.

Do not try to outrun a boar. Climb a tree or some boulders instead. If that is not an option, use last-minute side-steps.

To kill a boar, shoot or stab it in the following weak spots:

- Its face.
- Between its shoulder blades.

- Its belly.
- Just beneath its front legs.

Bulls

A bull is a non-castrated male cow. It can be aggressive towards humans.

When a bull stares at you, be quiet, don't move, and look for an escape route.

If he starts to charge, run to safety. In the event he is too close, remove a piece of clothing and wave it to the side to distract him. Remain still as he charges, then throw the clothing away from you. With luck, the bull will head for it. Run to safety.

Herd Animals

Species of animals that habitually stay in large social groups are herd animals. They may be wild or domestic. Common examples are cows, sheep, horses, and zebras.

Herd animals are generally not dangerous, but a stampede is life-threatening if you get caught in it. In that case, determine where the herd is headed and get out of the way. If you can't do that, run beside it. Do not lie down. Horses may run around you, but most other herd animals won't.

Mountain Lions

The mountain lion is a large cat species native to the Americas. It's also known by the names of cougar, puma, panther, and catamount.

Mountain lions are more active at dusk and dawn, and are more likely to attack individuals than groups of people.

Do not run from a mountain lion. Instead, make noise and back away slowly while waving your hands high in the air. Make yourself bigger by standing tall, opening your coat, putting a child on your shoulders, etc. If the mountain lion continues to approach or circle you, throw stones at it.

To fight off a mountain lion, protect your neck and throat, and hit it around the eyes and mouth.

Sharks

There are many species of sharks, and some of them are among the largest predatory fish in the sea.

Sharks don't normally eat humans, but they may "have a taste" if they're curious. Unfortunately, even a small nibble from a shark is life threatening.

They are most likely to attack during the twilight hours and in darkness.

Things that attract sharks include:

- Body fluids such as blood and urine.
- Bright colors.
- Schools of fish.
- Shiny objects (so remove your jewelry in the ocean).
- Trash.
- Silhouettes that resemble prey, such as those of bodyboarders.

When a shark approaches or circles you, that's a sign it will attack. Put your head underwater and shout to try and scare it off. If that doesn't work, hit it with whatever you have. Use your palm heel if you have nothing else. Aim for its eyes, gills, and nose.

Sharks are more likely to attack individuals, so if you're in a group, bunch together and face outwards.

Snakes

Most snakes will flee from humans, but may attack if threatened. When you see a snake, freeze and back away. Give it plenty of room to escape. If a python tries to constrict you, try to unwrap it by the head.

EARTHQUAKE

An earthquake occurs when there is a sudden slip on a fault line in the ground. Most are minor, but sometimes they are devastating.

Big earthquakes can also trigger other natural disasters, such as tsunamis.

If you live in an earthquake-prone area, do what you can to secure all movable and heavy objects.

In the case of an earthquake, get to the lowest point of a building, and/or to one of the following (in a rough order of preference):

- Inside a well-supported interior door frame.
- Under a large piece of furniture (hang onto it).
- An inside corner of the building.
- A hallway.

Stay away from:

- Unstable objects, including those on the floor above you.
- Elevators.
- Glass.
- Kitchens, tool sheds, etc.

If you're in bed, cover your face with a pillow.

If you're in a car, stay in it. Stop in an open area (not on or under a bridge) and crouch below seat level. Be extra careful on the road when you resume driving.

If you're outdoors, get to an open space. Lie flat on the ground and cover your head and neck. Do not go underground or in any tunnels, and get away from tall structures, including trees. If you're on a hill, the safest place is on top.

Beaches that are not below cliffs are safe during an earthquake, but should be evacuated as soon as the major tremors finish, in case a tidal wave occurs.

Once the earthquake is over:

- Listen to the media and prepare to evacuate as instructed.
- Turn off gas, electricity and water at the mains. Do not make any sparks or flames or use electricity until you are 100% sure there are no gas leaks. Do not try to turn the gas back on yourself. Let the gas company check for leaks and turn it back on.
- Be careful when opening cupboards.
- Do not take shelter in a damaged building. Build a temporary one from debris instead.
- If you are trapped underground, move slowly, and whistle or tap SOS. Do not yell if you need to conserve oxygen.

SOS

SOS is the universal sign of distress. Its pattern is:

... - - - ...

This can be transmitted as:

- short short short
- long long long
- short short short

When tapping SOS, use the length of pauses between the taps.

- Tap tap tap. Pause.
- Tap. Pause. Tap. Pause. Tap. Pause.
- Tap tap tap, pause

FIRES

The first sign of fire is the smell of smoke. If you smell smoke, investigate it. The sooner you confirm it, the better chance you have of escaping or extinguishing a fire.

Extinguishing a Fire

A fire dies without oxygen. Smother one with sand, fire blankets, and/or by using a fire extinguisher and other firefighting tools.

Be careful when using water. If it is an electrical or oil fire, water will make it worse. A better idea is to make an improvised fire extinguisher:

- Fill any bottle 3/4 full of baking soda.
- Pour vinegar into the bottle.
- Spray it on the fire.

Shaking up a can of beer and opening it in the direction of the fire also makes a good improvised fire extinguisher.

As soon as you feel that you cannot control a fire get out of danger and call for help.

To put out a living creature who is on fire, push him to the ground and smother him with a blanket. The heavier the fabric, the better. Get the person or animal to roll around. Cool any burned areas of skin with copious amounts of water immediately.

House Fires

An out-of-control fire in a home can be devastating, but is often easily avoided by taking some basic precautions:

- Have fire extinguishers, especially in the kitchen. Dry chemical extinguishers are good all-rounders. At the very least, have buckets of sand and water close to high risk areas, including outside.
- Install smoke alarms, preferably battery-operated ones with CO alarms built in). Use a minimum of one per level. Replace the batteries and test them biannually.
- Keep your chimney clean.
- Have a shovel, rake, and buckets handy in case of small outdoor fires.
- Clear all small trees and brush within 10m (30ft) of your home, and clean up all loose foliage. Rake up leaves and clean your gutters.
- Store flammable materials safely. Keep firewood at least 10m (30ft) from your home.
- Avoid burning anything outside on windy days/nights.
- Ensure all fires are completely out (cold to the touch) before abandoning them.
- If a fire is approaching, douse your house with water.

Do the following every evening before going to sleep:

- Make sure your electrical sockets are not overloaded.
- Switch off and unplug appliances.
- Extinguish candles.
- Don't leave flammable things on hot surfaces (such as clothes on heaters).

Teach everyone in your household:

- Where the fire extinguishers are and how to use them.
- When to abandon the fire fight.
- Multiple escape routes from each room, along with rally points.

- Individual responsibilities, such as calling the fire department and getting the pets.
- The home fire plan.

Here is an example of a home fire plan:

Shout "FIRE, FIRE, FIRE" and the location of the fire to let everyone in the house know.

Fight the fire if it's small enough. If you can't fight it:

- Call the fire department.
- Close the windows and doors, and turn off the lights to contain the fire.
- Grab your bug-out bag if it is close to you.
- Escape to the rally point. Stay low, and test door handles and other surfaces with the back of your hand before opening them.

Practice your fire plan every month.

When you are inside and the fire is outside, and there is no time to evacuate, stay inside. Close all the blinds and curtains, block the gaps around doors and windows, and stay away from outside walls.

Once the fire has passed, go outside and put out any small fires. Be careful of smoke inhalation.

Vehicle Fires

Keep a fire extinguisher somewhere you can reach it easily. Don't store it in the trunk.

When a vehicle is on fire, try to put it out, then exit the car. Even if you put it out, you need to ventilate the vehicle, because the fumes are toxic. If you can't put it out, distance yourself from it and call for help.

To move a vehicle on fire—to prevent it from burning down your house, for example—put it in a low or reverse gear and bounce the car out with short bursts of the ignition. It will jerk forward violently. Do not get in the car to do this.

Escaping Fires

There are various ways to escape fires. What you use depends on the situation.

In all cases, keep your clothes on for protection and soak them in water if you can. Remove all heat conductors (jewelry, electronics, etc.). A wet cloth over your mouth and nose will help with smoke inhalation.

The best thing to do is avoid the fire. Go around it if you can. When that isn't possible, head for any natural firebreak, such as water, a large clearing, a deep ravine, etc. Move downhill and/or into the wind. These are the directions that fire travels the slowest. Smoke is a good indicator of the direction of wind.

When in a building, study and take a picture of evacuation routes and emergency exits. Team up with others, and take the fire stairwell if you can. Never take the elevator. Mark your movements with Sharpies or post-it notes so rescuers can follow your path. Place them at or lower than knee height.

If you are in a vehicle, stay in it. Don't try to drive through thick smoke. Park in a clear area or pull off the road, but don't risk getting bogged down. Turn on your headlights, shut the windows, and seal the air vents.

Creating a fire break is a good option when the fire is some distance away and there is no way of avoiding it. The object is to burn off all the fuel in a particular area, so that the main fire has nothing to burn. The burnt area you create is the fire break, and is safe for you to stand in. To make a break, determine the wind direction carefully, then light a fire along as wide a line as possible. Make it at least 10m wide.

When you are desperate, you can try to run through a fire. If this is what you plan to do, then the sooner the better. Don't try it through thick vegetation. Use a hose to clear your path if you have one available.

As a last resort, bury yourself. Clear the area of foliage and dig as deep of a hollow as possible. Throw the dirt onto a coat, blanket, or something similar, then lie face down and pull the coat with the dirt on it overtop yourself. Cup your hands over your mouth and nose, and try to hold your breath as fire passes over.

FLOOD

A flood occurs when water covers land that is normally dry. Common causes are heavy rain, snow melt, and massive ocean waves.

Minor flooding is usually not a problem, but large floods can be quite destructive. The higher up you are, the safer you will be from flooding.

When you are caught in a flood, get inside a building. Prepare your survival kits and some kind of raft. Even a floating door is better than nothing. Turn off the gas, electricity, and water at the mains, then move to an upper floor or roof with a shelter. If the building has a sloping roof, tie everyone on.

Unless you live on the coast or are forced to evacuate, stay put.

Evacuating

To evacuate in a flood, seek shelter on the highest ground possible.

Do not attempt to cross water unless you're certain that it won't be higher than the center of your vehicle's wheels, or your knees if you're on foot. Even a small drop can make a big difference in water level. Be careful of a bridge already underwater. It may be missing.

If your car dies, abandon it.

After the Flood

Contamination of food and water is high after a flood, so stay away from floodwater and fresh food that has come into contact with floodwater.

You can eat canned food that has been exposed to floodwater. Wash the cans with soap and clean, hot water before opening them. Purify all water.

LANDSLIDE

(a.k.a. Mudslides, Debris Slides, etc.)

A landslide is like an avalanche, but with earth falling down a slope instead of snow.

Any occurrence that makes the slope unstable, such as a heavy rain or an earthquake, can cause a landslide. They mostly happen in past landslide areas and embankments along roadsides.

When you notice a landslide approaching, get out of the way. If you are in a building, get to a higher floor, if possible, and stay away from windows and any objects that could hurt you.

If you can't get out of the way, roll into a tight ball on the ground and protect your head.

Once the landslide is over, keep away from the slide area in case another one happens. Watch out for any other dangers it created, such as fallen power lines or gas leaks.

Repair and replant in damaged ground as soon as possible.

LIGHTNING

Lightning is a massive electrostatic discharge of energy in the atmosphere.

Your chance of surviving a lightning strike is high (90%), but you will probably suffer severe injury of some sort.

To limit the possibility of injury, stay indoors when there is a lightning/thunder storm. Avoid high ground and isolated tall objects.

Do not shelter in the mouth of a cave (deep inside is okay) or under an overhang of rock. If you plan on sheltering in a cave, be sure you have at least one meter of space around you.

If you are outside, find low, level ground. Crouch down on something for insulation (nothing metal or wet), and make yourself as compact as possible.

When you have nothing to use as insulation, lie flat.

Tingling skin and/or the sensation your hair is standing on end is a sign that lightning is going to strike close by. If you feel either one, drop down to the ground immediately.

PANDEMIC AND PLAGUES

A pandemic or plague is a widespread infectious disease, such as Ebola. It usually lasts about three months, and can have secondary waves.

To protect yourself from a pandemic, keep up to date with vaccinations and stay healthy in general to give yourself a stronger immune system.

Monitor the media for warnings and updates. If you fear a pandemic hitting, increase your stocks so they'll last three months or more (see part 2 of this book) and/or evacuate the area.

When the pandemic hits, avoid going out in public. If you have to, be very diligent with personal hygiene. Wear a mask, and avoid touching your face after touching public things, such as ATMs. See the *Biological Contamination* chapter for ways to make improvised masks.

Seek immediate medical attention for anyone who gets sick. In the case of a societal breakdown, quarantine sick individuals by sealing off an area with plastic sheeting and duct tape.

SANDSTORM

A sandstorm (or dust storm) is a wall of sand carried by the wind. It is common in arid and semi-arid regions.

When you are caught in a sandstorm, get to the downwind side of a natural shelter. Protect your eyes, nose, and mouth.

If you are traveling through the desert, sit or lie down in the direction you are traveling. Stay still until the storm is over.

When in a car, get off the road immediately. Turn off your headlights and turn on your hazard lights.

TORNADO

A tornado (or twister) is a destructive force of rotating air. It has a calm eye like that of a hurricane.

Tornadoes sound like loud spinning tops. If you hear (or see) one coming, take shelter below ground—in a basement, for example. If there is no suitable shelter, close all the doors and windows and get to the center of the lowest floor, preferably into a small room such as a bathroom or closet. When there is no small room, get under sturdy furniture or cover your body with a mattress.

Sheltering inside a car or caravan is a last resort. Any building is better. As you drive away, drive at right angles to the tornado's predicted path. Stay away from overpasses and bridges.

No matter where you are, if the tornado gets too close for you to escape, get as low as possible and protect your head. If you're outside, even lying down in a ditch is better than nothing.

QUICKSAND

Quicksand is a mixture of fine sand (or silt or clay) and water. It has a spongy texture.

It's impossible to be completely submerged in quicksand, but you will get stuck. Carrying a pole will help you get out.

When you fall in, remain calm and move slowly. The more you struggle, the more stuck you will get.

Lay your pole on the surface of the quicksand and use it to guide your back into a floating position, with your arms and legs spread out.

Shift the pole under your hips, at a right angle to your spine.

Pull your legs out, one after the other, and move to the nearest solid ground.

You can do this without a pole, but it will take longer. The main thing is to move slowly into a horizontal position.

Pulling someone out of quicksand is very difficult. Use a long stick or rope, and do it slowly.

TROPICAL CYCLONES

A tropical cyclone is a large, rotating storm. It is also called a hurricane or a typhoon.

During a tropical cyclone, there will be a period of calm. This is the eye of the storm, which may last up to an hour before the storm picks up again. In that time, prepare for flooding and be ready to evacuate if instructed to do so.

Indications of an oncoming tropical storm include:

- Abnormal barometric pressure variations.
- Dense cirrus clouds converging towards the storm.
- Increased ocean swell, especially if coupled with highly colored sunsets or sunrises.

Dense cirrus, often in the form of an anvil, being the remains of the upper parts of cumulonimbus.

When a tropical storm is approaching, you must decide whether to stay or evacuate.

Don't try to evacuate during a hurricane, only before it. Go as far inland as possible, and keep away from river banks.

If you choose to stay, keep indoors. Board up the windows and secure any outdoor objects that might be blown away. Shut off the gas and water at the mains, and bunker up in the centermost and lowest point of the building.

If you are in a vehicle during the storm, put on your seatbelt and cover your head and neck.

When in the great outdoors, try to find a cave. A ditch is the next best place, and failing that, the lee side (the side that shelters you from the wind) of any solid structure. If there is no shelter, lie flat on the ground.

During the eye, move to the other side of your windbreak or find better shelter. Other than that, moving during the storm is not advisable. If you have to, keep low.

Try to stay clear of things that may turn into flying debris, such as fences, coconuts, small trees, etc.

When at sea, batten down the hatches and stow all gear.

TSUNAMIS AND TIDAL WAVES

Tsunamis and tidal waves are technically different, but in the context of disaster survival, they are both giant waves.

When there is an earthquake, when you see a rapid fall or rise in shorelines, or when there is some other warning of a tsunami or tidal wave, evacuate immediately.

Seek high ground as far away from the shore as possible. Move at least 50m (150ft) above sea level and 4km (2.5mi) inland.

A tsunami may follow an earthquake. After it has passed, follow standard flood procedure.

VOLCANO

A volcano is a fissure in the earth that can spurt ash, gas, and lava. There are many shapes and sizes of volcanoes. The thing most people think of is a cone-shaped mountain.

Rumblings, steam, and other activity are signs that a volcano may erupt. There may also be acidic rain and a smell of sulphur from nearby water sources.

There are numerous hazards that come from a volcanic eruption.

Lava

Lava is flowing, molten rock with temperatures up to 1200 C.

You can outrun it, but it won't stop until it hits a valley or cools off. Once it cools, it becomes rock.

Missiles

The missiles from a volcano can be rock, molten lava, and other things. The best protection from them is to stay indoors and/or wear a hard hat.

Ash and Acid Rain

Ash and acid rain can affect places that lava and missiles won't reach. If you are outside while they're falling, wear a mask and goggles to protect yourself. Once you reach shelter, remove your clothing, wash any exposed skin well, and flush your eyes with clean water.

The ash will make roads slippery, so take care while driving.

Gas Balls

Take refuge in an underground shelter or under water while gas balls pass overhead.

Mud Flows

A mudflow is a landslide made predominantly of mud. It can occur during or after a volcanic eruption. See the *Landslides* chapter for what to do in case of a mudflow.

AIRPLANE HIJACKING

In the case of an airplane hijacking, you must size up the situation as best as possible. If it is likely you will get away safely, keep a low profile and appear compliant. When you feel that you will probably die anyway, you may as well try to take down the terrorists. You can apply the strategy below to any armed individual, such as a public shooter.

Disabling a Terrorist

If you aren't already, get into an aisle seat, but don't get seen making a swap.

Team up with at least one other capable person near you who is in an aisle seat, and communicate your plan, including who will do what and when or when not to act. Gather any improvised weapons, shields, and restraints you can without being noticed.

As a terrorist walks past, box him in. The person in front should grab his weapon while the others incapacitate and/or restrain him. Once he is down, use his weapon to take down the rest of the terrorists. If you need to reinforce his restraints, tie the reinforcements over his existing ones.

Lock the terrorists in the bathrooms and guard them.

BIOLOGICAL CONTAMINATION

Biological contamination is when harmful microorganisms (bacteria, viruses, fungi, and/or parasites) infect water, food, or the air. In a disaster scenario, this may be because of biological warfare and/or societal collapse, as opposed to an isolated case of contaminated food.

Signs of biological contamination include:

- Aircraft dropping objects or spraying, especially enemy aircraft in wartime.
- Breakable containers or unusual bombs, particularly those bursting with little or no blast and/or muffled explosions.
- Sick or dying people, animals, and/or vegetation.
- Smoke or mist of unknown origin.
- Strange taste in food or water.
- Tears, difficult breathing, choking, itching, coughing, dizziness, etc.
- Unusual substances on the ground, vegetation, or your skin.

When you think there are biological contaminants in the air, cover your mouth, nose, eyes, and skin. Use protective equipment if available (bio-suit and gas mask). Stay out of depressions (cellars, ditches, valleys, etc.) and tall vegetation.

Move crosswind or upwind to higher ground.

Cook all food and boil all non-bottled drinking water.

When you can't evacuate, take shelter. If you need to construct a shelter, do it in a clear area away from vegetation, and place the entrance at a 90° angle from the wind.

Improvised Protective Equipment

Here are some ways to improvise protective equipment:

- Soak a clean cloth in one tablespoon of baking soda mixed with one cup of water. Cover your nose and mouth with it.
- A sponge soaked in clean water makes a decent air filter.
- Use swimming goggles to protect your eyes.
- Cover your skin in dry powder, such as flour or fine cornstarch, to block your pores.

In case of improvised dust masks and clothing, the finer the weave the better. Silk is a good option.

Finding Food

A big issue when it comes to biological contamination is finding uncontaminated food. Don't trust anything that has been exposed.

Canned food, sealed bottled water, and other packaged goods should be okay. Wash them in clean water before opening them.

If you want to hunt, take the following precautions:

- Avoid animals that appear to be sick or dying.
- Skin an animal carefully to avoid contaminating the meat.
- Leave 0.5cm (0.2in) of the meat on the bone and discard the bone.
- Discard all internal organs.
- Cook it very well.
- Avoid aquatic food sources, egg shells (an egg itself is safe), and milk straight from an animal.

For plants, the order of safest to eat is as follows (safest first):

- Grows underground (potatoes, carrots, etc.).

- Can be peeled (oranges, bananas, apples, etc.).
- Everything else.

Regardless of what you eat, wash it well and remove the skin or outer layer. If in doubt, boil it for 10 minutes.

Decontamination

The information in this section applies to instances of biological, chemical, and nuclear contamination.

Decontaminate yourself and all equipment as soon as it is safe to do so. Do this before entering living spaces, preferably outside and downwind from your home and others.

If you are indoors, choose a room with running water and seal it off. The lower you are in the house, the better.

Self-decontaminate if possible.

If you are helping young children, the elderly, etc., you must wear protective clothing and self-decontaminate afterwards.

If your government has an official decontamination plan, follow it. If not:

- Cut off all clothing. Don't pull it over your head.
- Place eyewear in bleach.
- Put all clothing in a trash bag and seal it.
- Rinse (don't bathe) your body in cold water (warm water opens pores). Start from your head and work your way down. Do not scrub.
- Wash your hair with soap and water while bending forward to protect your face.
- Wash your hands and face with soap and water.
- Blot your body with a cloth containing soap and water.
- Flush your eyes with copious amounts of water.

- Clean your teeth, gums, tongue and the roof of your mouth frequently.
- Rinse, gargle, and spit if you're able.
- Put on fresh clothing.

When washing, it is best to use a neutral soap, such as unscented Castile soap, to prevent any chemical reactions.

For trouble areas, you can:

- Blot with 0.5% bleach solution. Don't do this on your face.
- Use a 5% baking soda/water mix to reduce irritants, such as riot-control gas.
- Put flour or talcum powder on the affected area. Wait 30 seconds and then wipe it off.

Bury what you can. Wash everything else with a 5% bleach solution.

Seek medical attention when it's safe to do so, even if you think you weren't affected.

CHEMICAL ATTACK

A chemical attack is almost the same as a biological attack, but with chemicals instead of biological contaminants.

Some signs to look for are:

- Acid rain.
- Aircraft spraying, especially enemy aircraft in wartime.
- Droplets of oily film on surfaces.
- Liquid vapors.
- Low clouds, but no rain.
- People falling ill and/or collapsing.
- Small animals and/or vegetation dying.

To combat chemical attacks, take all the same actions as for biological contamination.

In addition, put as many walls between you and any toxic cloud as possible. Try to get inside a building. Once inside, shut and lock all doors and windows leading to the outside. Turn off air circulation and seek refuge in a room that:

- Has few entry points.
- Is high up.
- Provides access to water, food, and a phone.

Cover all gaps (outlets, vents, etc.) with plastic sheeting/duct tape. Place wet towels under doors.

Do not drink from the tap.

EXPLOSIONS

This advice is for "small" bombings, such as those caused by grenades or pipe bombs. For instructions on what to do in case of nuclear, biological, or chemical attacks, see the related chapters.

When an explosion occurs, expect a second one. Cover your mouth, take shallow breaths, and evacuate the area.

If you see a grenade (or a similar explosive device):

- Take cover if the cover is within three steps.
- If not, leap away from the grenade and onto the ground.
- Adopt the blast position.

When indoors, use the earthquake action plan during the bombing. After the bombing, wait a minute before leaving, then do so as quickly as possible and get away from the building. Refer to the *Fire* chapter for information about escaping buildings.

If the bomb goes off on the street, get inside the nearest building. Failing that, get to a clear spot away from the blast area. Stay away from cars or other suspicious things that could be additional bombs.

The Blast Position

Lie on your stomach, with your feet facing the grenade. Cross your legs. Open your mouth a little to prevent your lungs rupturing. Cover your ears and keep your elbows pressed tightly against your ribs.

MARTIAL LAW

Martial law is when the military takes control of policing. It may be introduced in the case of civil unrest or a foreign invasion.

The diminishment of civilian rights usually occurs over time. As soon as you see the first signs, start to prepare by increasing and hiding stockpiles.

When martial law is enacted, expect the following:

- Curfews.
- Checkpoints.
- Home searches.
- Restrictions on items needed for daily living (energy, food, medicine, guns, etc.).
- Community roundups.

In the case of civil unrest due to disaster, it is generally okay to wait it out. Once order is restored, things will start to go back to normal.

If you think it might be a while, you can leave for a few months—to go visit a relative in another state, for example—until things settle down.

When martial law is due to a new government, strongly consider leaving, especially if there are signs of radical propaganda.

Surviving Martial Law

To survive martial law, be the gray man. Keep to yourself, and don't attract any attention.

Carry your ID at all times and follow (or at least appear to follow) all imposed regulations.

Avoid encounters with authority at roadblocks or public gatherings. When confronted by authority, be polite and comply (within reason) without volunteering any important information.

NUCLEAR ATTACK

Nuclear warfare can create destruction on a massive scale in a short time and have long-lasting effects. There may not be time to evacuate, and if there is, it will be pandemonium.

If you can evacuate, do it. Go as far away as possible and prepare to be away for the long term—that is, for months to years. Leave behind and/or bury any contaminated clothing and equipment. Travel upwind from the blast and take routes that won't be congested by everyone else.

If you're not evacuating, building a fallout shelter is your best chance of survival. This option takes more time and money than most people are willing to spare.

Other than that, take the same action as for a chemical attack.

When you're outdoors with no shelter available, dig a trench/foxhole at least 1m (3ft) underground and stay there. Cover the trench with material if you have it.

Adopt the blast position (see *Explosions* chapter).

After a blast, your survival under nuclear conditions depends on your time of exposure, distance from the source, and shielding.

When in a secure shelter, stay there for at least 24 hours, and preferably a few weeks if you have the supplies. If you must leave (to procure water, for example) do not expose yourself for more than 30 minutes at a time, and follow the decontamination procedure upon return (see the *Biological Contamination* chapter).

POWER OUTAGE

A power outage is when your electricity goes out. It may affect just your property or occur on a wider scale, and can happen for any number of reasons. One of the most common causes is bad weather.

To prevent a power outage, avoid overloading your electrical grid. In the case of an outage affecting more than your property, there isn't much you can do to stop it, but you can do things to prepare for it.

Install night lights in every room that will come on automatically during a power outage.

Keep your freezer filled with PET bottles (soda bottles) filled with ice. When filling them with water, leave room for the expansion caused by freezing.

Create an emergency lighting kit that contains:

- Flashlights.
- Matches and/or lighters.
- Spare batteries, light bulbs, and fuses.
- Long-burning candles.
- Some form of entertainment, such as cards.

Consider buying a portable back-up generator.

When a power outage occurs, take the following action:

- Get your emergency lighting kit.
- Ensure the safety of family members.
- Turn off all electrical appliances to prevent power surges when the power returns.
- Check if the power is out in the whole neighborhood. If not, the problem is probably your main fuse. Be careful when checking it, in case the outage is a trap.

- Monitor the local media.
- Stay together.

To conserve the life of your food, don't open the fridge or freezer. In outages lasting two or more hours, pack perishable goods into a cooler with ice. You can refreeze food if it has ice crystals and stays under 4°C (40°F).

Down Power Line

When you come across a downed power line, do not touch it or anything it is or might have touched. Get away from it and call the power company to fix it.

Electrocution

When someone is being electrocuted, use non-conductive material, such as a dry piece of wood or plastic, to separate them from the source of electricity. Gloves will not give you protection. Once you're both safe from the source, apply first aid.

PLANE CRASH

There isn't much you can do to prevent a plane crash, but to give yourself the best chance of survival:

- Choose an aisle seat within five rows of an exit door.
- Keep your shoes on and your survival kit in your pocket, especially for the first and last 10 minutes of the flight.
- Count the number of seats to the exits in front and behind you and take note of any obstacles.
- Check that your flotation device is where it should be. If it's missing, inform the flight crew.
- Read the safety card (every single time!).
- Pay attention to the safety demonstration.
- Keep your safety belt on during the entire flight.
- Do as instructed by flight personnel.

Landing a Plane

When the pilot is incapacitated and there is no one else able, you may need to land the plane. Go to the controls and use the radio to get the landing instructions, which will vary depending on the type of plane you're in:

- Put on the headset if there is one.
- Check the steering wheel or instrument panel for the talk button.
- Press the button and use the international distress call of "Mayday! Mayday!"
- Give your situation, destination and plane call numbers, which should be located on the top of the instrument panel.
- Let go of the talk button and listen for a response.
- If there is no response, try again.

- Try 3 to 5 times, waiting 10 seconds for a response between each attempt.
- If there is still no response, tune the radio to 121.5, which is the international emergency channel, and try again.

Once you have made contact with someone, follow their instructions to land.

If no one answers your distress signal, you will need to try and land the plane unguided. Here is a quick description of the main controls in a plane to help orient you.

Yolk. This is the steering wheel. It has the same effect as in a car, but is much more sensitive. It also allows you to control pitch. Pull back to pull up and push forward to dive.

To fly steadily, keep the nose of the plane about 8cm (3in) below the horizon and the wings even.

Altimeter. This is the red dial on the instrument panel. It indicates your altitude.

The small hand shows your height above sea level in thousand-foot increments. The large hand shows the same in hundreds.

Compass. The instrument with a small plane on it. The direction the nose of this plane is pointed in is the one you're going.

Speedometer. A plane's speedometer usually measures the places speed in knots. Cruising speed is 120 knots. Below 70 knots, you may stall.

Throttle. The throttle controls thrust. Pull it towards you to slow the plane and descend. Push it away to speed the plane up and ascend.

Fuel Gauge. This is usually on the lower part of the panel.

Landing Gear. If the plane has a retractable landing gear, there will be another lever between the seats near the throttle. It will be shaped like a tire.

Some planes have a fixed landing gear, so they will not have this lever.

Ground Pedals. Use the ground pedals when on the ground.

- The upper ones are the brakes.
- The lower ones control the direction of the nose wheel.
- The right pedal will move the plane right.
- The left one will move the plane left.

To land the plane, first find the longest and smoothest place you can to land. Use the yolk to steer towards it, and circle around it if you need time to land properly.

When you're ready to land, slow down to 90 knots by pulling back on the throttle.

Let the nose drop to about 11cm (4in) below the horizon then deploy the landing gear (if applicable), unless you're landing on water.

If you have enough fuel, fly over to look for obstructions, then circle back to land. Give yourself a wide berth.

Line up the landing strip so that it's just off the right-wing tip at 1,000 ft.

As you approach land, pull back on the throttle. Do not let the nose drop more than 15cm (6in) below the horizon.

The rear wheels should touch first, preferably at about 60 knots (stall speed). Pull all the way back on the throttle, ensuring the nose doesn't dip too steeply.

Gently pull back on the yolk as the plane touches the ground. Use the pedals on the floor to steer and brake.

If you are headed into an obstruction (e.g., trees), let the wings take the impact.

Once you have stopped, get everyone out as soon as possible using the plane's emergency exit procedure.

SHOOTINGS

In a shooting scenario, you have three choices: Run, hide, or fight.

As soon as you hear shots, crouch down so you're a less obvious target. Identify the direction of the bullets and the proximity of the shooter(s) so you can decide what to do. If you decide to move right away, remember that ricocheting bullets move along solid surfaces. To keep yourself safe, leave a small gap when following walls, and squat or crawl when getting low.

Anyone displaying the following characteristics may be a shooter:

- Acting nervous and/or suspicious.
- Wearing long-sleeved shirts or long pants in warm weather.
- Making adjustments and/or checking placement of the weapon.
- Has a shape or an unusual bulge in his clothing that you can visually identify as a gun.

Run away from the shooter if you have the distance and cover to get to. When you do:

- Turn a corner as fast as possible.
- Zigzag from cover to cover.
- Do not leave your cover position until you know your next one.
- Keep an eye on the shooter if possible.

When you can't run, hide, preferably inside a room with few entry points.

- Lock and barricade all entry points, close the blinds/curtains, turn off the lights, and get behind the piece of cover the furthest away from entry points.

- Put your devices on silent and call for help.
- Place something in the window so rescuers can identify your location, but only if doing so will not give away your location to the shooter(s).
- Do not open the door for anyone you can't positively identify. The shooter(s) may be tricking you.

Fighting a shooter is a last resort. To do it effectively, work in a team. Have one person go for the shooter's weapon and the other for his legs. Use any improvised weapons you can.

When there's a shooter outside and you're inside, secure all the entry points and hide inside the building. To lock a door with a closer arm:

- Get a belt or anything similar.
- Wrap it tightly around the closer arm at the point closest to the door.
- Secure it tightly.

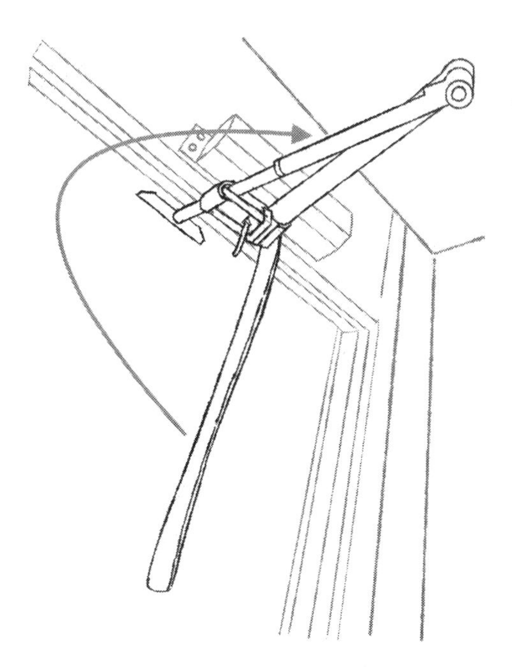

Cover vs Concealment

Concealment is anything between you and your enemy that hides you from sight, such as a desk.

Cover will hide you from sight too, but will also stop bullets. A concrete pylon is an example of cover. The more powerful the gun (or blast), the thicker the cover needs to be.

Concealment is better than nothing, but always aim to get behind cover.

When you are not the primary victim, or in the case of a random shooter, rapid concealment will prevent you from becoming a target. However, you still want at least some cover to protect you from stray bullets. If need be, hide behind a car tire or in a gutter

In the case that you are a target, cover is paramount.

RIOT

The best way not to get involved in a riot is to avoid large crowds, especially political gatherings and protests.

If you're out and about and a riot starts, get and stay inside until it is over. Secure all the entry points and stay away from the windows. Contact the authorities for instructions.

Leave if there is a fire or rioters penetrate the building. If possible, exit to a quiet street.

If you get caught up in a riot, don't get between opposing sides, such as protesters and police.

Blend in with the crowd and walk with it as you make your way to a safer area, such as:

- The sides of the crowd.
- The back of a group of peaceful protestors
- High ground.
- Inside or to the back of a building.

Avoid:

- Front lines.
- Barriers.
- Law enforcement (they will be defensive).
- Public transportation, especially below ground.

If you're exposed to riot-control agents, make an improvised mask and decontaminate yourself as soon as possible, following the steps in the *Biological Contamination* chapter.

When in a vehicle during a riot, never drive towards crowds or police lines. Avoid high-traffic areas and major roads, and don't stop until you're safe. If people block you, honk and carefully drive through or around them.

Surviving a Stampede

Don't rush to the exit if you're in a stadium environment. That's where the stampede will head.

Adopt a steady stance, keep your arms up to protect yourself, and use a shuffle step to move forward. Fill in any gaps and keep moving.

If you get knocked off your feet and can't get back up, get on your knees, bend over, and protect your head.

PREPPING 101

STOCKPILING

Stockpiling is a major concept in prepping, and is the easiest way to get through a temporary collapse scenario. This chapter will discuss some common factors to do with stockpiling, some of which cross over to prepping in general. Subsequent chapters in this book discuss stockpiling for each prepping scenario in more detail.

What to Stockpile

Stockpile everything you need in order to live, and some things you don't.

Other chapters in this section go into detail about what and how to stockpile. They cover food, water, energy (fuel, batteries, etc.), health, hygiene, and more.

Stockpile things that you normally use, especially perishables, when it's possible to do so. When building a stockpile, don't forget about the special needs of others in your family, including any infants, pets, elderly, and disabled people.

It's always a good idea to keep a bit of cash in case the power grid goes down. In long-lasting situations, it may become useless, but in the short term it will be useful for purchasing emergency supplies if credit cards and ATMs stop working.

Where to Store Your Stock

For those who own larger properties, this won't be much of an issue.

If you have a smaller living space, you have to get a bit more creative, but you'll be surprised how much "dead space" there is under beds and inside closets. You can also create shelving to make use of wall space.

There are other things to consider besides space, such as shelf life and access. You'll need to store some things under certain conditions. You'll also need to have easier access to perishables so you can rotate them as needed.

Finally, you'll need to hide some (or most) of your stocks, in case the government or looters want to take them. Leave some poorly hidden as decoys, and keep more secret caches on and/or off your property. You can also create secret rooms, hide stuff in the walls, etc.

It's a good idea to store similar things together for organization. If you have many different areas, keep a list of where you put what. For example, you might put canned foods in the kitchen pantry, and spare batteries and lighting supplies under the bed.

Do not keep records of any hidden stocks.

When to Start Stockpiling

Now!

Start to build up your supplies now. Do it little by little, and before you know it, you'll have enough to last you through a mini-collapse scenario, e.g., a few weeks after a natural disaster.

If you notice signs of disaster and/or long-term collapse, enlarge your stocks with emergency stockpiling (explained later in this chapter). Signs of various disasters are covered in part 1 of this book. Signs of long-term collapse include increases in:

- Violence.
- Crime.
- Police presence.
- Propaganda.
- Economic trouble.
- Prices.

And decreases in:

- Available resources.
- Citizen rights.

Take Small Bites

Prepping is a massive subject, and the concept of stockpiling so much is overwhelming, but don't let that prevent you from starting.

As with all big goals, the best way to tackle it is by breaking it up into smaller pieces. Here are some tips:

Prep for what's most likely to happen in your area.

Do a little bit regularly instead of a lot all at once. For example, buy $10 worth of extra food a week as opposed to three months' worth in one hit.

Use the rule of three to decide how much long you want to prep for:

- Three days to start. You probably have about a week's worth right now.
- Three weeks is good. In most disaster scenarios, basic infrastructure will be restored within three weeks.
- Three months is ideal. In a long-term situation, three months gives you time to figure out how to become self-sustaining.
- Three years or more if you're a hardcore prepper. The longer you are prepared for, the better.

Write a list of what you need to get for the timeframe you are prepping for, and build up supplies in each area evenly. This will prevent you from overstocking in any one area. Having a year's supply of batteries, for example, is great, but not if you only have food for a week.

Rotation

Rotation will prevent your perishable stocks from going bad. Food is the main concern, but this practice also applies to medicines, fuel, batteries, etc.

Use the "first in, first out" concept. This means that whatever you bought the longest ago will get used and replaced by the latest things you buy.

This will also force you to store things you actually use, which is what you want.

Emergency Stockpiling and Scavenging

When you notice signs of collapse or a natural disaster approaching, you need to increase your supplies. The earlier you do this, the better. You want to beat the crowds, both for safety and so you can get what you need before things run out.

Your primary targets are:

- Long-lasting food.
- Batteries.
- Drinking water.
- Health and hygiene supplies.
- Lighters/matches.

You will also want to keep an eye out for these things after the initial collapse.

Next, focus on heavy-duty maintenance supplies and tools that you don't need electricity to use (start stockpiling these beforehand as well). These are things like:

- An axe.
- Gardening shears.

- A hand drill.
- A hand saw and sawhorses.
- A hammer and framing nails.
- A screwdriver and three-inch galvanized steel deck screws.
- A staple gun and half-inch staples.
- Tarps (10mm thick, 10 x 12 inch and 6 x 8 inch, with sturdy eyelets).
- Plastic sheeting.
- Plywood (3/4 inch thick).
- Quick clamps.
- Ladder (tall enough to reach your roof).
- Duct tape and electrical tape.
- Gorilla glue, liquid nails, and a caulking gun.
- 3-in-1 oil/WD40.
- Rope and cordage (550 mil-spec paracord).
- Personal protective equipment (gloves, safety glasses, etc.).

Go to the following places (choose the ones closest to your home) to buy or find your emergency stocks:

- Warehouses.
- Storerooms.
- Distribution and industrial areas.
- Non-obvious medical clinics, such as plastic surgery clinics, eye clinics, dentists' offices, and veterinary clinics.
- Restaurants and fast-food stops.
- Auto and truck repair shops.
- Repair shops (television, vacuum, computer).
- Entertainment complexes (movie theaters, skating rinks, stadiums, etc.).
- Strip malls.
- Maintenance sheds.
- Farm/livestock supply houses.
- Abandoned homes.

Avoid:

- Big-box stores (e.g., Costco).
- Super-centers (e.g., Walmart).
- Mom-and-pop stores that are not abandoned.
- Armories.
- Bases and government installations.
- Weapon stores.
- Hospitals.
- Places with armed security.
- Anywhere there are large crowds.

In general, avoid places where people have weapons.

Buy what you need in the early stages of collapse. Once police response is ineffective, you will need to resort to scavenging. Scavenging is not the same as stealing. Take what others have left behind, but do not take from private residences whose owners still live in them.

There will be other people scavenging as well, and they may try to steal from you. Take the necessary security precautions. For example, work in groups, with appointed security.

Conservation

Even if you expect the collapse scenario to be short-lived or for it to end soon, you must start and continue to conserve (ration) your supplies. You never know how long a bad situation will last, no matter what others tell you.

First, try to get what you can by other methods (scavenging, foraging, etc.) and use your stocks to supplement what you find.

Use funnels to avoid spilling liquids. Label them by use and keep them separated to avoid cross-contamination. You don't want to accidentally use your kerosene funnel for water, for example.

Self-Sustainability

No matter how much you stockpile, your resources will eventually run out. If you want to survive in a long-term collapse scenario, you'll have to become self-sustainable. Hopefully, your stockpiles will be large enough to see you though until you can do that.

But you don't need to (nor should you) wait for a collapse in order to start. The chapters to follow will give you ideas on how to work towards self-sustainability. For instructional projects, please visit:

www.SFNonfictionBooks.com/DIY-Sustainable-Home-Projects

BUGGING IN/OUT AND RALLY POINTS

Bugging in is getting home and waiting out a disaster. Bugging out is evacuating. You need to have plans for both, and set guidelines for when to use each one.

For most families, the default plan will be to bug in. Once everyone is accounted for, you can access the situation and decide whether to stay or bug out.

You also need to have rally points set so you have alternative places to meet if you need to bug out without going home first.

Create a set of emergency family text codewords to let each other know which rally point to meet at. Make them easy to commit to memory, and don't tell anyone else what they are. You could also use these code words over the phone, but texts are more likely to get through in a grid-down situation.

It's also a good idea to leave some kind of sign out in front of or near your home in case the message doesn't get through and someone returns. That way, they'll know where to find you without having to enter the house.

Bugging In

Plan multiple routes to get from places you frequent—such as school/work, favorite hangouts (mall, bowling club, gym, etc.), and the local supermarket—to your home.

For each route, figure out the safest (which is not necessarily the quickest) way to get home via car, public transport, and on foot.

If your plan is to bug in and stay there (during a riot, for example), do not leave the house unless absolutely necessary. Nine times out of 10, it's safer to defend your home than it is to protect yourself out on the street.

Bugging Out

When bugging in (or staying bugged in) is too dangerous, you'll need to bug out.

First, you need to designate a bug out location. Where that is exactly will on your resources and where you live. It may be a holiday home in another state, an abandoned factory on the other side of town, a relative's house, a motel, a campground, etc. As a general rule, avoid refugee camps, public shelters, or trying to survive in the wild.

It is also a good idea to have at least two bug-out locations: one close to home (within a day's hike/a short drive), and one further out.

Regardless of where your chosen location is, it needs to be safe and stocked with at least a few days' supplies, and preferably a few weeks' worth. In the same way as you would when bugging in, you need to plan several routes from your home and other frequently visited places to that location.

When you bug out from your home, shut off your services (water, gas, electric) and head off in the early hours of the morning (1-5am). This is the safest time, with the least traffic. Be discreet about your departure, and keep the radio on at low volume, so you can get updates and avoid trouble areas.

Rally Points

A rally point is any place you have arranged to meet up. Default rally points are places like your home or bug-out locations. Good secondary rally points are safe places (malls, gas stations, police stations, etc.) that are relatively convenient to get to from the places you and your family frequent. They can also be spots where you've stashed caches.

It's a good idea to stash low-risk caches at your secondary rally points. These caches should contain food, water, flashlights, poncho, matches, etc. Avoid including anything expensive or dangerous, so if a cache gets found, it's no big deal. Check these small caches once a month.

Finally, have a plan for how to meet at rally points if you are separated. For example, you might instruct family members to wait at a given point for up to 60 minutes at sunset, and leave a marker to show they were there.

You can (and should) also set temporary rally points for if people get lost or split up when out and about (such as when they're at the mall or traveling). In places you frequent, such as the local supermarket, make the rally point the same place each time.

When you are traveling, a prominent landmark works well, because getting directions there will be easier. Make sure you specify exactly where and visit it if possible. For example, train your family members to meet at the water fountain in the town square, on the side facing City Hall.

Creating Routes

You can create and save routes to and from bug-in/-out locations using Google Maps, and/or with a pencil on a standard street map. Create ones that avoid danger and high-traffic zones (people, funnels, road blocks, high-crime areas, main roads, city evacuation routes, etc.) while not straying too far out of the way.

Once you have at least three alternate routes, you need to prioritize and label them. For example, you might call one "school-to-home route 1 (SH1)." This will allow for easy (and covert) communication about which way you plan to go, as well as make it easier for people to find each other should the need arise.

Commit the routes to memory and practice driving, walking, and taking public transport along them during the day and night (if safe to do so). Take note of anything useful, such as how long each one takes, useful places along the way (food, water, shelter, etc.), sketchy neighborhoods it passes through, and so on.

Keep yourself up to date on all important places and developments in the area, and adjust your plans as needed. For example, note new roadworks, building sites, hospitals, pharmacies, police departments, or water sources.

BUG OUT BAGS (BOBS)

A bug-out bag (BOB) is a single bag of supplies you can quickly grab and go when needed. It's basically a survival kit with at least several days of provisions. It must have the ability to provide you with water, food, shelter/warmth, fire, rescue, health, and security. Many items in it will be of a general nature, but when you pack it, you should also consider likely events in your area. This way, no matter what the emergency, you can grab your BOB (if it is safe to do so) and bug out.

Everyone in your household, including your pets, should have their own BOB, and they should keep it somewhere easy to access in case of an emergency. Under the bed or next to the nightstand are good options.

Assign responsibility for pets, infants, etc. and their BOBs. Do it now, so there is no confusion when an emergency arises.

What to Put in Your BOB

The exact contents of your bag will depend on what you're comfortable using and what events you feel are most likely to happen. You can also add some personal and/or comfort items if you have the room and weight tolerance (you may have to carry it all day, every day). The bag itself must be comfortable and sturdy.

Once you have put your BOB together, ensure you rotate the perishables every few months.

Here is a list of items to consider including in your BOB:

- Cash (small bills).
- Knife (steel).
- Multitool.
- One liter of water (minimum).
- Water filter (portable/hiking style).

- Food (long-lasting and ready to eat; think energy bars, trail mix, multivitamins and electrolyte mixes).
- A spare set of clothing.
- Emergency blanket.
- Poncho (transparent white is best).
- Lighters.
- Ferro rod.
- Flashlight (headlamp).
- Whistle.
- Shortwave radio with AM/FM (battery-operated and compact).
- Batteries.
- GPS-capable cell phone (with SIM card and charger; a cheap "burner" phone is ideal).
- Maps.
- Compass.
- First aid kit (with antibiotics).
- Toiletries (essentials).
- Sewing kit.
- Duct tape.
- Paracord (5m).
- Weapon and ammo (if legal).
- Notebook and pens/pencils.
- Plastic bags.
- Photocopies of important documents (see end of this chapter).
- Swimming goggles.
- P100 mask with an air vent.
- Special-needs items.

For infants:

- Food/formula.
- Water.
- Clothes.

- Comfort toys/blankets.

For pets:

- Food.
- Water.
- Leash.
- Toy.

It is a good idea to get a cage for your pet and train him/her to sleep in it. That way, it will be comfortable for him/her to stay in when you need to leave in a hurry. Keep his/her BOB on top of the cage.

CACHES

A cache is a hidden store of supplies.

You may have caches in your home, at rally points, along your routes to bug-out locations, or anywhere else you think makes sense.

You can also have different caches for different things, either to separate items or to pack them for specific scenarios.

Containers

The container you choose for your cache must protect the items you are storing. It must be waterproof, airtight, and corrosion-resistant. Other characteristics to consider depend on how easy to access it needs to be and where you're going to hide it. For example, is it suitable for burial?

A PVC pipe with sealed ends is a popular option, as it is durable, inexpensive, and easy to waterproof, but any other durable box will work as long as you seal it properly. If it has a rubber lining, that will make your job easier. Test the seals by submerging the cache in hot water and looking for bubbles.

Additional Protection

Waterproof the individual items before putting them in the waterproof cache. You can use heavy-duty trash bags, vacuum sealing, plastic sheeting and duct tape, etc. Before sealing the items in, add desiccants and remove as much air as you can.

Adding desiccants will absorb extra moisture. Silica gel packets are common and cheap. Use 5g for every 3.5L (1gal) of space. If in doubt, add more.

There are many other choices for desiccants, which may or may not work as well. These include rice, salt, zeolites, calcium sulfate, and kitty litter.

Hiding Your Cache

A big factor in deciding where to hide your cache is accessibility. You need to be able to access it in an emergency, as well as for maintenance.

Another factor is concealment. Put the cache somewhere that is not obvious, but that is easy for you to relocate. Burying your cache is a good option, especially if it is off your property. If you need semi-regular access, consider a shallow bury—for example, place it in a small depression under a large rock.

When the cache is on your property, you can hide it in your walls or roof.

Other options include hiding it at your workplace, in a storage container, in a PO box, on a rooftop, or even underwater (if you have a boat moored at the local harbor, for instance).

Some places to avoid include:

- Private property that's not yours (unless you are paying for it, in which case, keep anonymous if possible and never miss a payment).
- Populated places (parks, beaches, vehicle access roads).
- Abandoned buildings.
- Anywhere with security cameras.
- Places that may be developed in the future (outside urban areas).

The way you store your cache will also determine its location. For example, if you're burying it, you'll want to avoid choosing ground that contains obstructions, such as rocks, large tree roots, or pipes.

You'll also want to avoid ground that is high in moisture or prone to rain run-off. In general, don't bury it in lowlands.

Wherever you choose, you need to scout the location out before actually putting your cache there. Decide on a possible area from home first using Google Maps/Earth. Then go out there to assess it further. Check out exactly where you think you will stash/bury your cache, as well as how secure the area is.

You will need to get the cache and tools out there and have enough time to stash (or bury) it without anyone seeing you. Stake it out at different times, too, in case there is a change in activity level on weekends vs weekdays, or at night vs during the day.

Once you have an exact location, you need to remember where it is. Perhaps you can remember without a prompt, but I wouldn't rely on that alone unless you have a photographic memory. Things (especially your memories) change over time. A better idea is to write non-specific instructions that you understand, but that will be useless to others. Other options are to store the location in your GPS, record the grid references on a map, and/or include a small Bluetooth tracker in the cache.

Keeping the Secret

There is no point hiding a cache if other people know about it. In fact, don't even tell anyone you plan to do it. If you live in a rural area where word spreads easily, purchase supplies in from a different town.

When physically hiding (or accessing) your cache, you need to be as covert as possible. Do it at dusk or dawn on a Sunday or Monday, and wear gloves so there are no fingerprints. Use a flashlight only if you need to, and make sure it's red or blue (never use white light). Ensure you leave no signs of your presence. This means parking your car out of the way and hiking in without making an obvious trail. Unless you're burying your cache, you also need to consider ways to camouflage it.

Ensure no GPS devices (phones, cars, etc.) record where you're going and have a cover story in case anyone comes along. For example, say you're doing a time capsule project or treasure-hunting with a metal detector. Take equipment to confirm your cover, and ensure you have food and water.

If you need to access your cache, take the same precautions. Always use a different path in/out (to prevent making trails) and minimize access to it. The more often you access your cache, the less secure it is. To improve security, you can also create decoys and/or misdirections by burying a layer of trash above the cache.

Car Supplies

You can store additional supplies in your car. Keep them in the trunk for security, except for the last two items, which you'll need to have handy in case of an emergency.

- Blankets.
- Additional food, water, flashlights, and batteries.
- Fuel.
- Recovery and repair supplies.
- Entertainment (books, cards, laptops, etc.).
- Chargers.
- A small fire extinguisher.
- A glass-breaker.

Do not put your personal BOBs in the trunk. Keep them within reach in case you need to leave your car in a hurry.

Important Documentation

Gather all the following documentation. Keep the originals in a fireproof safe (or some other secure place) and tell your family its location. Photocopy everything and keep the photocopies in your BOB. Ensure everything is kept current.

- Your will.
- Your powers of attorney.
- Emergency/important contacts (numbers and addresses).
- Your passport (or other ID if you don't have one).
- Insurance information.
- Proof of residence (utility bill).
- Access to finances (do not keep a photocopy of this in your BOB).
- Personal info sheet and recording.

A personal info sheet is a single sheet that will aid rescuers in finding and/or identifying you. Each family member should handwrite their own info sheet and make an audio recording of the information. This is so rescuers will have writing and voice samples.

Each sheet/recording should include the following:

- Name.
- Nicknames.
- Place of birth.
- Date of birth.
- Address.
- Phone number.
- Physical Description (including specific identifiers like tattoos or birthmarks).
- Prescriptions (eyes, medication).
- Instructions for prescriptions.

- Vehicle (color, type, license plate number).
- School/work address and contacts.
- The contact details of closest friends/relatives.
- Hobbies.
- Education.

FOOD

When the power goes out and your fridge turns off, you have limited time until things will begin to spoil.

To conserve the cold environment, don't open your fridge or freezer unless you need to, and eat the things that will spoil the fastest first. If you have advance warning, turn your refrigerator and freezer down to their coldest settings.

Depending on how much food you have, you may not have to open the freezer for 24+ hours. If you keep it packed with ice-block bottles, it will stay cold enough to keep the food inside edible for at least 48 hours.

Stockpiling

If you are stockpiling supplies for a set amount of time, say three weeks, you'll have to figure out how much you need to store.

You may already have a good idea of how much your family consumes, but what you stockpile is likely to be different, since you will not stock perishables. Remember to take that into account.

The foods you stock should have the following characteristics:

- Easy to prepare (minimal cooking needed).
- High in calories and nutrients.
- Long shelf life.
- Things you like to eat (or at least can tolerate).

When considering stocking a certain food that you don't usually eat, test it out a few times before buying it in bulk.

Start with a base of staples (wheat, rice, beans, fats/oils), and supplement them with canned/packaged foods. Basic spices and condiments are a good idea, and so are multivitamins.

Mylar bags are good for long-term storage of staples (wheat, rice, beans, flour, pasta, etc.). Fill each one with a single food type, throw in some oxygen absorbers (do not use desiccants with food), squeeze out as much air as you can, and then seal it. Store the mylar bag of food inside a sealed bucket, and you're done. Depending on what you're storing, this will keep your food edible for a long time. White rice, for example, can keep for 10+ years.

Here is a video of the process:

https://www.youtube.com/watch?v=fk9bodAtJ80

Store your buckets in a cool, dry spot and throw some mothballs around the storage area.

Self-Sustainability

Producing and preserving your own food is the best way to ensure a constant supply of perishable foods. Start doing/learning it now, so that when the time comes, you already have a good system in place. As an added bonus, it's healthier and cheaper than store-bought food.

Some options to consider are:

- Growing a garden.
- Raising chickens (for eggs and/or meat).
- Foraging.
- Hunting.

When you produce more than you need, you'll want to preserve it. Consider:

- Air-drying.
- Canning.

- Dehydrating.
- Fermenting.
- Freeze-drying.
- Pickling.
- Storing food in a root cellar.
- Salting.
- Smoking.
- Using a solar-powered fridge.

Cooking

If you currently cook with electricity, you will be stuck as soon as the power goes out (unless you have a backup power supply).

If you use gas, then you may last a little longer, but eventually that will run out.

A propane camping stove works well for short-term situations. You can stockpile the fuel cans, and the stoves' simple mechanics mean they are easy to fix.

In the long term, pairing solar cooking (solar ovens are easy to make) and a mud oven or rocket stove is a good system. Use solar when you can, and your stove as a backup.

A thermal cooker (also easy to construct) is a good way to conserve fuel. Heat the food up whichever way you choose, and then let it cook slowly in the thermal cooker.

There's another technique that's not really cooking, but that you can use to increase the nutritional value of grains and beans: sprouting them. To do this:

- Soak them in clean water for 24 hours.
- Drain them and place them in a jar.
- In a few days, they will sprout.

WATER

The average person needs about four liters (one-gallon) of water a day to cover their drinking, cooking, and sanitation needs.

Stockpiling

Keeping clear, two-liter water bottles in the freezer is a good start. This is a source of clean drinking water, will help keep the freezer colder for longer, and you can reuse the bottles for SODIS purification.

Two-liter PET soda bottles are what you want to use. Clean them well before refilling them with drinking water. Other types (such as milk or juice bottles) may be less durable and/or cause bacterial growth, while if you use anything bigger than two liters, SODIS may not work properly. Anything is okay.

Larger containers, like water bricks, are good for bulk storage, especially if you can stack them.

When the disaster is predicted, fill up your bathtubs and other containers with water. Your bathtubs will leak over time. Prevent this with a large bathtub water bladder (a WaterBOB). The WaterBOB will also keep the water cleaner than your bathtub, since it is sealed.

Scavenging

Before dipping into your stores, get what water you can from around your home. Doing this early will prevent the water from getting too contaminated or evaporating. Here are some ideas for water sources:

- Water pipes. Open the spigot at the highest point in your home, and drain the water from the lowest spigot.

- Hot water tank. Turn off the gas or electricity, drain the tank, and filter out any sediment.
- Toilet tank (not the bowl). Only use this if you haven't added any chemical cleansers to the tank.
- Recreational sources (swimming pool, hot tub, or pond).
- Garden hoses/pipes.
- Anywhere else water accumulates.

All these water sources are okay to use for hygiene purposes. If you plan to drink the water from any of them, make sure you purify it first.

Self-Sustainability

The easiest way for most people to achieve self-sustainability in their water supply is to collect rainwater. Several IBC containers plumbed together may provide enough room to store all you can during a downpour and last you until the next one. Larger, purpose-built water storage tanks are available if you have the space and money.

Digging a well is a little more work, but it's worth the effort to have a more stable water source.

Relying on local sources (rivers, lakes, ponds, etc.) is an option too, but in that case you will need to go and collect the water every day. Unless the source is right next to your home, it is a hassle. It can also be dangerous in a collapse scenario, when contamination (from dead animals, for example) is more of a risk.

If you live in the appropriate climate, you can consider fog harvesting.

A combination of a well and rainwater collection is doable for most people. Together, these sources should supply enough water for a family to live off without too much restriction.

Purifying Water

Purify all water before consuming it. Even tap water is unsafe to drink after a disaster.

Gravity or inline filters are reliable, and you can stockpile them.

You can also use SODIS, which involves using UV rays to sterilize the water. Fill up a clear two-liter PET bottle with water and leave it in the sun for six hours or more. On cloudy days, leave it out for 48 hours or more. On rainy days, this technique is ineffective.

Another option is to boil water for five minutes or longer.

Finally, you can use chemicals, such as bleach, purification tablets, or iodine. Out of all these methods, this is the least preferred since it uses chemicals, but it will make the water safe to drink and you can purify large amounts inexpensively depending on what you use.

Bleach is probably the cheapest method. Store plain bleach (no additives or scents). A 5.25% calcium hypochlorite solution is good. Use two drops per every liter of water, and shake/stir it well. Wait for at least 60 minutes before consuming the water.

ENERGY

Energy plays a big part in our modern lifestyles, and many people would not last long without it.

This chapter discusses things we use that rely on energy/electricity to run, including lighting, heating, cooling, communication, and entertainment.

Batteries

Solar-powered appliances are good but not 100% reliable (you need the sun), and hand-cranked appliances aren't worth the effort. Batteries are reliable and work when you need them to, assuming they haven't run out.

Stock enough batteries to run all your important stuff three times over. Store them in a dry, cool/room temperature place out of direct sunlight. Don't store them in equipment (except obvious cases, such as in a bedroom flashlight), as they will slowly trickle energy and can cause corrosion.

Standardizing your equipment so it all takes the same type of batteries makes things a lot easier, and means you can mix and match in an emergency. Choose a common type of battery, like AA or AAA.

Rechargeable batteries are good, but remember that you need electricity to charge them. A solar charger is a good way to do that when the grid is down.

Generators

A 4KW generator is enough to run most small appliances directly from the generator. You need 10+ KW to run an average-sized household

Generators come with a few downsides. The noise they make makes you a target for thieves. They also need fuel, engine oil, and maintenance to keep them going.

If you decide to get a generator for emergency use, start it at least once a month and keep it somewhere well ventilated.

Vehicle Inverter

You can use your vehicle as an improvised generator for small appliances, such as a portable fridge or a laptop.

Attach an 800-watt inverter directly to your car battery with jumper cables, then plug your device into the inverter.

Smaller inverters can also power small appliances depending on their consumption.

You don't even need the car to do this, just the DC battery. However, running the car keeps the battery charged (but consumes fuel).

Fuel

There are many types of fuels you can stockpile depending on your needs. Store as much of the type you need as you can within safety and legal limits.

Use fuel stabilizers where applicable, and adopt a dating and rotating system for them.

Some fuels to consider are:

- Gasoline.
- Diesel.
- Propane.
- Biofuels.
- Charcoal.
- Gel fuel.

- Paper bricks.
- Firewood.

Lighting

Keep the following things together in a "lighting kit":

- Flashlights (mag lights, headlamps) and spare batteries.
- Candles (slow-burning "survival" candles)
- Lighters/matches.

Solar patio lights will charge during the day, so you can bring them in at night.

In unexpected outages, night lights plugged into the wall sockets will come on when the power fails. This will allow you to see well enough to grab your lighting kit.

You can improvise oil lamps using anything that's fireproof and has a small depression, such as a ceramic saucer. Don't use metal, as it will burn you. Pour a little oil (or anything that is slow-burning) and add a string wick.

If you're outdoors, wrap cloth around a long stick and soak it in flammable liquid to make an improvised torch.

Heating

The best way to heat a home without consuming too much energy is to start with good insulation, especially in the floor. Install weather stripping and thermal curtains (thick blankets will work).

In a well-insulated home, if you cook inside, the heat generated may be enough to keep the house warm throughout the night, depending on your climate.

A few thermal solar panels work well if you get enough sunlight.

If that is not enough, heat the person using layered clothing and blankets, as opposed to trying to heat the whole house.

If you will use a heater of some sort, choose one room to heat and have everyone gather in it. Ensure it is ventilated.

Some heating options include:

- Fireplaces.
- A wood stove.
- Battery-operated heaters.
- A rocket stove.
- Gel fuel.

Cooling

The basis of efficient cooling is similar to that of heating. Use cooling curtains (survival blankets with the shiny side facing out, for example) and ventilation.

Cook outdoors.

Cool the person using manual, solar, or battery-operated fans and water.

Communication

Keeping in touch with each other and the outside world is important for long-term survival.

Small battery-operated shortwave radios that can receive AM/FM are cheap and use very little energy. Tune into local and national radio stations every other day to listen for any useful information.

Cell phone emergency alerts are available in most countries and will warn you of any natural disasters.

When you and your family members are contacting each other during times of disaster, text messages are more likely to get through.

Once the grid is down, you can use walkie-talkies tuned into personal radio stations. They're not secure and their range is short, but they're easy to use. You can switch channels regularly if you are worried about security. Let everyone know beforehand how often and which channels to switch to.

Get more information here:

https://en.wikipedia.org/wiki/Personal_radio_service

Other radio services (such as, GMRS, MURS, or HAM) require you to have a license and/or training.

To access the internet in off-grid scenarios, you need to rely on cell phone signal, satellite internet, or wireless internet service provider (WISP). In a collapse scenario, it's unlikely these things will work either.

As long as you have a phone signal and a data plan, you can turn your phone into a hotspot, buy a dedicated hotspot, or use the internet directly from your phone.

If you live in an area where the cell phone signal is poor, you can use a signal booster (cell phone repeater). For it to work, you need at least a little reception and electricity.

Satellite internet or WISP are your final options. Both are expensive for what you get, but may be the only things available.

Entertainment

In a grid-down scenario, there will be enough work to keep you busy during the day. For recreation you will need to go old-school. Think cards,, board games, books, sports, etc.

Self-Sustainability

To go 100% off-grid is possible, but it takes some work and/or money.

Going partially off-grid is fairly easy. You can buy ready-made solar-powered appliances such as lights, heating panels, and fans.

For complete energy setups, you need to look into photovoltaic solar systems, wind turbines, hydroelectricity or geothermal systems.

Solar is the most accessible option for most people, but it isn't cheap to set up.

If you have running water on your property, then a hydroelectric system is a relatively cheap and reliable way to go.

Wind turbines can work if you live in a windy area.

Geothermal power can work almost anywhere, but is very costly to set up.

Most people can create a self-sufficient power supply by combining two or more of these methods.

HEALTH AND HYGIENE

Poor health and hygiene can quickly lead to illness, and in an austere situation, even the common cold can lead to death.

Stockpiling

At a minimum, stockpile the following:

- Bleach.
- Anti-bacterial soap.
- Toothbrushes, toothpaste, and dental floss.
- Condoms.
- Toilet paper.
- Sanitary products.
- First aid supplies.
- Personal aids and medications (glasses, diabetes medication, etc.)
- Antibiotics (consider veterinary meds as a substitute).
- Heavy-duty garbage bags.
- Multivitamins.

Toilets

When the water pipes stop flowing, you will feel it first in the family toilet. You can flush it manually by pouring a bucket of water directly into the bowl. When water is scarce, you will want to switch to a bag-and-dispose method:

- Empty as much water out of the bowl as you can.
- Line it with two garbage bags.
- Put some sand and bleach over the waste.
- Dispose of it after several uses.

You can use this same method with a bucket. This is good for portability, in case you need to move the family toilet outside.

Create a hand-washing station and ensure everyone maintains impeccable hygiene, especially after using the bathroom and before handling food.

Trench Toilet

For anything more than a few days, dig a trench toilet outside. Choose a suitable location:

- Not too far from the house.
- Downhill and away from food and water (at least 60m/200ft).
- Not near rainwater runoff (put it on raised ground).

Dig the trench 1.5m (5ft) deep and 0.5m (2ft) wide. A trench 1m long (3ft) is a good length for a small family.

Create a privacy barrier around it.

You will need to squat to use the toilet, or build it up if you want to sit. Keep a bucket of dirt and a small shovel next to it, and throw some in after each use. Cover it with an improvised lid, such as a piece of plywood, when it isn't in use.

Self-Sustainable Toilets

A trench toilet is sufficient for most people. If you want a bit more normalcy, you can construct a composting outhouse or have a septic tank installed.

Both of these are larger projects, and it will be much easier to install them before a collapse scenario occurs.

Showers

Washing regularly is important. If water is low, take a bird bath once a day and restrict a full wash to once a week. For a bird bath, you only need a cloth with some soapy water. Use it to clean your face, armpits, and crotch—in that order.

When water is really low, you can downgrade to an air bath. Expose naked skin to the sun and air for at least an hour. Be careful not to get sunburnt.

This is also good for clothes. Give them a good shake first, then hang them inside out in the sun.

Hot Water

Hot water for washing is a luxury, but can be incredible for morale.

The simple way to do it is boil some water and mix it with room-temperature water until you get the desired temperature.

Solar water heaters work well as long as you have sun. You can buy portable camping ones or make your own from black poly pipe. The water can get too hot, so mix it with room-temperature water if needed.

An upgraded version of a solar shower is a batch heater, which you can construct from an old water heater tank. This will keep a larger amount of water heated, and you can plumb it right into your pipes.

Soap

Learning to make soap from scratch is a valuable skill. The process isn't very hard once you learn how, but it can be dangerous to work with lye.

In a nutshell, lye (hardwood ash and water) + fat (animal or vegetable oil) = soap. Of course, there is more to it than that, so if you plan to do it, follow some detailed instructions.

You can replicate this process to make an improvised dishwashing soap when cooking with a wood fire or oven. Mix wood ash with the fat from cooking and a little hot water. Do not use this on your body.

Growing soap plants is another good option.

General Waste

Reuse as much as you can. For instance, be sure to wash out cans.

Use the "bash, burn, and bury" method for whatever you do end up throwing away. Dig a pit away from food, water, etc. and downwind from your house. Compact your trash as much as you can and throw it in the pit. Burn it, and then cover it up.

Medicine

In a collapse situation, you won't be able to rely on hospitals, ambulances, pharmacies, or any other medical services. Learn basic medical skills now, while you still can.

A wilderness first aid (WFA) course will equip you with the knowledge of how to handle most medical emergencies. Going more in-depth than a WFA is great, but at the very least, you need to know basic first aid.

Knowing some proven home remedies will also be invaluable.

It is important to fix any injuries as soon as possible, no matter how small they are. In austere situations, small things can become big problems fast.

For more information on wilderness medicine check out *Wilderness and Travel Medicine*:

www.SurvivalFitnessPlan.com/Wilderness-Travel-Medicine

Prevention

Besides practicing good hygiene, the best way to prevent injury and illness is to be healthy.

Build a strong body and immune system now, while times are good, because it may turn into survival of the fittest when the grid goes down.

The path to good health is not a secret, but here are some key points anyway:

- Have a daily exercise routine.
- Eat a balanced diet that includes less refined sugar and more vegetables.
- Avoid drugs, alcohol, and smoking. This includes pharmaceuticals, except the ones, like insulin, that you might need in order to live.
- Get a thorough medical (including dental) at least once a year.
- Fix health issues permanently, if possible. Lasik surgery will eliminate the need for eyeglasses, for example.
- Keep your immunizations up to date.

SECURITY

When society breaks down, people will start doing things they normally wouldn't in order to survive.

Keep safe by increasing your passive home security, as well as by being prepared to actively defend your family and belongings.

Don't Be a Target

Keep your prepping actions a secret from anyone outside of your household, and even from them if they are not responsible enough to keep the secret (young kids, for example).

When things start to go bad, actively make your home unattractive to would-be scavengers and thieves. Minimize signs of food, water, and energy. If your home is the only one with lights and/or the smell of cooking, you will get visitors.

If the situation calls for it, consider creating signs of disease and/or death, such as quarantine signs.

Physical Security

Having decent security now will make it easier for you to enhance it in troubled times. It will also make you less of a target for everyday criminals. Here are some things to consider:

- Solid doors.
- Deadbolts.
- Motion-sensor lighting.
- Clearing blind spots in your front and back yard
- Reinforced windows and sliding doors.
- A nightly security routine (checking locks, etc.).
- A guard dog.

- An alarm system.
- A neighborhood watch.
- Hiding spots for valuables, in the walls or other hard-to-find locations.
- A safe room.

When the times comes, increase your physical security by:

- Keeping a guard shift at all times.
- Fortifying entry and exit points with sandbags.
- Creating fire positions.
- Covering your windows with plywood
- Building obstacles, such as wire entanglements, to funnel intruders to your line of fire.
- Setting up booby traps and early-warning systems.

No matter what changes you make to your home security, ensure you will be able to escape in case of a fire or other internal emergency.

Firearms

Having a firearm will dramatically increase your ability to defend your property. However, it will be useless and dangerous without the proper training.

Take professional classes on gun safety, how to shoot, and how to maintain your weapon. Don't forget to stockpile ammunition.

A 12-gauge pump-action shotgun is a great option for home security, and you can also use it to hunt. Choose a Remington 870 or Mossberg 500 with buckshot ammunition. Both types are reliable and relatively inexpensive.

Some accessories to consider are:

- Mounted flashlight.

- Pistol grip.
- Side saddle (for quick access to ammunition).
- Sling (especially if patrolling).

You can also consider handguns (which are easy to carry and conceal) and rifles (which are better for hunting).

Whatever you choose, stick with well-known manufacturers.

Body armor is another thing to consider.

Here are some basic gun safety guidelines:

- Always assume a gun is loaded.
- Never point it at something you don't want to kill/injure.
- Keep your finger off the trigger until you want to fire.
- Know what is getting shot (confirm your target and account for what is behind it).

Safety in Numbers

It is easier to fight off attacks if you join forces with your neighbors. Even whole towns can band together to keep unwanted drifters out.

Working together will also make it easier to achieve self-sustainability. Use the strengths of others to the benefit of all—for example, to create community gardens, ensure medical assistance, or build alternative energy projects. People with no specific skills can still help with important grunt work such as guard duty, general labor, or water collection.

Even though you are working together, it doesn't mean you need to share your family stores. However, it's a good idea to help out where you can, so others don't turn on you.

Bartering

When you need or want things you don't have, and money is no longer worth anything, you'll need to barter.

This can be a dangerous endeavor, especially if the word gets out that you have an abundance of something people need, such as fuel, food, water, or medicine.

To combat this, barter anonymously through a trusted third party or a respected member of the community, like a priest or doctor.

FURTHER EDUCATION

What you've learned about prepping from this book is only the tip of the iceberg.

It provides a good overview, but it is possible to delve into each skill/subject in greater depth. Depending on how much you want to learn, you could go from reading specialized books to taking online courses to studying for degrees.

For most people, being a jack of all trades is the best way to go. It's useful to learn the foundations of:

- Survival skills.
- Construction.
- Plumbing.
- Security.
- Mechanics.
- Food production and preservation.
- Electronics.
- Renewable energy.
- Water harvesting.
- Medicine (scientific and alternative).
- Whatever else you can think of that will be useful.

Do an Off-Grid Test

The best way to figure out which skills you should learn is to test run living off-grid.

Confine yourself to your property for at least four days. Shut off the gas, electricity, water, and all other services you rely on, and live as if you are the only one left on earth.

This will also highlight any deficiencies you have in your stores or sustainability plans.

Dear Reader,

Thank you for reading *The Disaster Survival Handbook.*

If you enjoyed this book, please leave a review where you bought it. It helps more than most people think.

Claim your bonuses:

www.SurvivalFitnessPlan.com/Book-Bonuses

Connect with like-minded people and discuss anything SFP related via the SFP Facebook group:

www.Facebook.com/groups/SurvivalFitnessPlan

A list of resources used in the creation of the Escape, Evasion, and Survival Series is available at:

www.SurvivalFitnessPlan.com/Escape-Evasion-Survival-Series

Thanks again for your support,

Sam Fury, Author.

AUTHOR RECOMMENDATIONS

Teach Yourself Escape and Evasion Tactics

Discover the skills you need to evade and escape capture, because you never know when they will save your life.

Get it now.

www.SurvivalFitnessPlan.com/Evading-Escaping-Capture

SURVIVAL FITNESS PLAN TRAINING MANUALS

Survival Fitness

When in danger, you have two options: fight or flight.

This series contains training manuals on the best methods of flight. Together with self-defense, you can train in them for general health and fitness.

- **Parkour.** All the parkour skills you need to overcome obstacles in your path.
- **Climbing.** Focusing on essential bouldering techniques.
- **Riding.** Essential mountain-bike riding techniques. Go as fast as possible in the safest manner.
- **Swimming.** Swimming for endurance and/or speed using the most efficient strokes.

It also has books covering general health and wellness, such as yoga and meditation.

www.SurvivalFitnessPlan.com/Survival-Fitness-Series

Self-Defense

The Self-Defense Series has volumes on some of the martial arts used as a base in SFP self-defense.

It also contains the SFP self-defense training manuals. SFP Self-Defense is an efficient and effective form of minimalist self-defense.

www.SurvivalFitnessPlan.com/Self-Defense-Series

Escape, Evasion, and Survival

SFP escape, evasion, and survival (EES) focuses on keeping you alive using minimal resources. Subjects covered include:

- **Disaster Survival.** How to prepare for and react in the case of disaster and/or societal collapse.
- **Escape and Evasion.** The ability to escape capture and hide from your enemy.
- **Urban and Wilderness Survival.** Being able to live off the land in all terrains.
- **Emergency Roping.** Basic climbing skills and improvised roping techniques.
- **Water Rescue.** Life-saving water skills based on surf life-saving and military training course competencies.
- **Wilderness First Aid.** Modern medicine for use in emergency situations.

www.SurvivalFitnessPlan.com/Escape-Evasion-Survival-Series

ABOUT THE AUTHOR

Sam Fury has had a passion for survival, evasion, resistance, and escape (SERE) training since he was a young boy growing up in Australia.

This led him to years of training and career experience in related subjects, including martial arts, military training, survival skills, outdoor sports, and sustainable living.

These days, Sam spends his time refining existing skills, gaining new skills, and sharing what he learns via the Survival Fitness Plan website.

www.SurvivalFitnessPlan.com

amazon.com/author/samfury
facebook.com/SurvivalFitnessPlan
twitter.com/Survival_Fitnes
pinterest.com/survivalfitnes
goodreads.com/SamFury
bookbub.com/authors/sam-fury

Printed in Great Britain
by Amazon

58882340R00066